DATA-DRIVEN DECISION-MAKING

ANDREW BARWIS

BEAVER'S
POND
PRESS

Edited by Kerry Stapley
Illustrated by Valerie Valdivia
Book design and typesetting by James Monroe

ISBN 13: 978-1-64343-623-4

Library of Congress Catalog Number: 2023913351

Printed in the United States of America
First Printing: 2023
27 26 25 24 23 5 4 3 2 1

Cover and interior design by James Monroe Design, LLC.

Beaver's Pond Press, Inc.
939 Seventh Street West
Saint Paul, Minnesota 55102
(952) 829-8818
www.BeaversPondPress.com

To order, visit www.datapurgatory.com. Reseller discounts available. Contact Andrew Barwis at www.datapurgatory.com for speaking engagements, book club discussions, and interviews.

For Gail

INTRODUCTION

It had been a long week. I had a project that never seemed to move forward despite five talented people working diligently on it for months. We'd been trying to optimize how we managed our server deployments, but the solution was ever elusive. There were too many variables to be able to solve all at once; I was starting to worry the coffee pot wouldn't handle the strain, and neither would we.

Over the past year, the team had come together under Jan, a leader who was tenacious, inquisitive, and blunt. He was the kind of person who would keep a meeting going long past the scheduled end point, asking question after question until he was satisfied.

His protégé was Mac. Mac was still learning the ropes, as he had come to us fresh out of school the year prior. His eagerness for learning led him to be easily molded by Jan's blunt honesty and insightful critiques. His career was just getting off the ground, but he had the world ahead of him.

Elise had started this year and was the team's junior member. She was undaunted by tasks that should have been beyond her capabilities and was learning and growing very quickly. I had underestimated her at the start and thought of her more as a support than as a full team member. She hadn't been discouraged by this—quite the contrary, she had automated her whole workload so it was done by ten o'clock each morning.

Val was between Mac and Jan in terms of seniority. She was harder for me to work with. On paper, she had the experience, and she had been successful on other assignments. But she didn't have confidence enough to stand up to Jan, and she seemed intimidated by Elise's fearlessness. I felt there was something untapped within her, but I didn't have the skills to help her express it.

K.T. rounded out the team. She brought decades of experience and a quiet competence that occasionally swerved towards aloofness. She and Jan would bicker when things got heated. I'd heard a lot of bickering lately.

Long after everyone else had gone home for dinner, I remained at my desk, staring at a draft email to my team. The subject line read "Project Canceled." I'd been working up the gumption to call the whole thing off. It was clear we were spinning our wheels. We either didn't have the right people or didn't have the right tools to grapple with the task at hand. But the project was so important to the company. Millions of dollars were at stake, not to mention people's livelihoods. Canceling was a bad option, but if we couldn't find a way to reduce costs, we just didn't have the money to pay our engineers. My email explained all of this in stark detail. I don't like to leave anything to the imagination. I just didn't have the heart to send it. Not yet.

Rubbing my eyes, I flipped back to my unread emails. Six more had come in since I'd last checked. Three of them were from Jan, each asking detailed clarifications to the answers I'd provided a dozen times before. Two of the emails were from the CTO, the first asking for a status update, the second demanding one. The last was a reply to an email I'd sent on a whim three weeks prior. I opened it. The others could wait.

"Andrew," it read. "Please join us for a tour. We will expect you tomorrow at 8:00 a.m." The email was signed "Virgil," and included the name and address of his institution. My initial email—containing the details of my predicament and a plea for help, advice, mentoring, or anything else he could offer—was attached below. I didn't know Virgil, and he didn't owe me anything. I had sent my email in desperation and had not expected a response.

Virgil had founded an institution for problem-solving. It had begun accepting students five years ago. Ever since the first graduate re-emerged from its vault of a campus, it had become legendary for producing candidates of such talent and skill that the world clamored for what was behind those locked doors—the secret to the institution's success.

No journalist had ever made it past the front desk. Virgil didn't respond to requests for comment or hold press conferences. His institution was an informational black hole, and I hadn't been able to see past the event horizon.

Despite the nonreligious nature of the school, it was structured as a cloister. Once a student matriculated, they had no contact with the outside world until graduation. No one could be seen milling about the sparse campus. This created a blank canvas onto which pundits painted all manner of rumors and conspiracy theories.

None of them were true, of course, and in retrospect, most were rather silly. The truth is always more mundane, yet more complex than any conspiracy story.

I cleared my calendar for the next day, sent my team an out-of-office notification, and got a fitful night's sleep. The anticipation of possibly getting help for my problems kept me running what-if scenarios long into the night.

The lockdown the institution had on proprietary information was extreme. The only two administrators who anyone had been able to link to the venture were Olivia Marshall, an enigmatic data scientist at the pinnacle of her career, and Virgil Benson, an expert in brain-computer interfacing. No one who hadn't experienced the institute knew how they were able to achieve their incredible results. I'd heard rumors that billions had been invested in their organization, but how those dollars were being used and who had backed such a venture was a mystery. I wasn't convinced there was any way I would learn enough, quickly enough, for the institute to benefit my project and my team, but I also didn't have any other prospects for a Hail Mary.

I woke up before my alarm, nervous energy powering

my morning routine. I gulped down breakfast and chased it with two cups of coffee, threw my notebook and pen in my bag, and hustled to the car. I caught a view of my face in the rearview mirror and realized, in my rush, I'd forgotten to shave.

I punched the address into my map. The institute was 28.4 miles away; my ETA was 8:04 a.m. I took a deep breath and peeled out of the driveway. I'd waited years for this and wasn't going to miss out because of a little scruff.

The drive took me out of the city, deep into rolling suburban hills. The farther I got from the city, the faster I was able to push it. The cloister was conspicuous as soon as I came up on it. Its sleek modern building was set back behind a sturdy black fence. The clock on my dash glowed 7:59. A pudgy guard checked my ID and pointed me toward a small guest parking lot.

I would have had my pick of spaces in the empty lot, except that there was a lanky older gentleman in a well-tailored linen suit standing on the sidewalk. I did a double take. I recognized this person because I'd seen his image alongside countless articles. It was Virgil. I parked next to him and jumped out of the car, leaving my bag in my haste.

"You must be Andrew. I'm Virgil," he said, extending a hand in greeting.

"Hi, I'm Andrew," I responded, grasping his hand. "It's great to meet you." I tried to start coolly, but the weight of my expectations opened the floodgates of my mouth. "I mentioned it in my email, but I'm in a tough spot and need your help. My team is in a bad place, and if I can't come up with a solution, I'm going to have to let them all go. I've followed you for years but never thought you'd respond to my email. Thank you so much!" The stress and anxiety of the project, along with my caffeinated nerves, pushed me to speak faster and faster. "I've got good people—talented people—that I'm about to lose."

Virgil put a comforting hand on my shoulder and said, "Why don't you come inside, and we can talk."

Virgil motioned toward the plate-glass front door. We walked as words continued to tumble out of my mouth.

"If I'm being honest, I messaged you out of pure frustration and never in a million years thought you'd reply. I know your time is valuable, and I don't feel quite right using you as a sounding board for this, but if anyone can help, it's probably going to be you or one of your students."

Virgil stopped and interrupted me. "You just laid out a lot. These are the kinds of problems our students solve in the workplace, and it sounds like you're under a deadline. That's a hard place to be. I recognize that, and I'm sure you have plenty of questions to ask. However, I didn't invite you here just to help solve your problem, if I can even do that. Our time is precious, so I ask that you trust me and accompany me on a journey. If, at the end, you still have any questions, you may ask them then."

This stopped me in my tracks. I stood a moment,

considering the offer. I'd barely laid out my issues to Virgil. There were nuances, externalities that he didn't yet appreciate. We had shifting priorities from the organization. The team had continually failed to deliver, and we had interpersonal issues. Every hour brought me closer to the end of my job. I was here due to his generosity, though, and I didn't want my impatience to shut down our discussion. In the end, patience won out. We turned to walk back up the corridor and into a massive hall.

A high, vaulted ceiling encased row after row of cylindrical machines strung out into long chains. It reminded me of a train station, albeit one that contained rows and rows of MRI machines. The room thrummed with the beat of large cooling systems buried beneath us. I peered inside one of the machines and was shocked to find a prone figure inside. At first, I thought the machine contained a corpse, but then I saw the heart-monitoring system that detailed each crest and trough of life in ten-millisecond intervals.

A few scattered caregivers attended to these half-human, half-machine digital centaurs, defraying the inevitable bits of entropy that, if allowed to accumulate, could pose real problems. A tear wiped here; a sensor adjusted there; they kept things in check. My own scruff was far outpaced by the faces of the capsuled men. One of the men we passed looked like he could have been captain of the *Pequod,* and the one next to him would have given Blackbeard some competition. After allowing me to take in the scene, Virgil began to explain.

"Each unit contains a student on a journey, much like the one we are about to take. The machine keeps its occupant in stasis while simulating a new reality for them. Each journey differs according to the needs and skills of the student. Some take as little as a month or two to complete the training, while others need more time. If you're willing, I'd like to offer you a guided glimpse into what our students experience. We will cover the breadth of their experience in the span of a day, but you'll only catch a glimpse of what they endure."

The offer was remarkable, and having come this far, I quickly agreed. I didn't have a week, let alone a year, but I could spare a day. He walked me to an empty machine. On its gleaming white exterior, it bore a small yellow inscription above an outline of a boat. "Heavenly Helmsman."

Virgil saw me staring at the logo. "That's a subsidiary of ours. We produce all our own machines off designs I perfected only recently." I lay back in the machine. A caregiver came over, handed me a lozenge, and instructed me to take it under my tongue. There was a burst of strong citrus flavor, and then my senses dissolved into nothingness.

CHAPTER ONE

I awoke to a gentle touch on my shoulder. I cracked one bleary eye and smacked my lips against the overpowering taste of orange. There were palm leaves and coconuts above me. They were framed by a cloudless blue sky.

As I sat up, my stomach turned. I tried to stand, but I knew my knees would buckle if I pushed it. Virgil hooked his arm beneath mine, and with our combined efforts I was able to stand and lean against a tree trunk as I took in my surroundings. The rows of machines were gone. We were standing in the high shade of a copse of palm trees. Spread out before us was a narrow sandy beach with a blue-green lagoon beyond it. Behind us, a tall peak loomed on the horizon.

"Welcome to the island," Virgil said. "Those feelings of anxiety, exhaustion, and shock will wear off soon. The summit of that peak," he said, gesturing, "is our destination. To reach it in time, we should be on our way."

My heart was racing, and my stomach was in knots. I took a tentative step around the tree trunk to see if my knees would hold. I took a second step and didn't fall. I let go of the tree and tried again, feeling more confident. I didn't want to waste time, nor look weak in front of Virgil. "I'll muddle through."

We departed, following a well-trampled trail that led away from the beach. As we walked, my senses slowly returned to me. My steps, which were initially wobbly and short, began to lengthen. I was able to stand straighter, and the exercise seemed to help clear my head. My initial shock of being completely transported gave way, and I began to feel calmer and more contemplative. The landscape, which seemed at first to be a wash of green, became more detailed. There were various species of grasses. I heard a rustling in the brush and glimpsed a fat brown rodent scurrying away. The fidelity of the simulation was incredible; it was clear that a lot had gone into building this world.

Shortly after we started up the path, we encountered a pair of travelers. The two men were picking their way up the trail, their backs bent under the weight of heavy packs. The man on the left was shorter and older. His head was balding, and gray hair grew unchecked at his temples. The younger man to the right wore clothes that were sweat soaked from his neck to his knees. Their full packs were lashed haphazardly. All manner of tools, tomes, and other instruments dangled from their packs. I hesitated to approach the men. They walked with a begrudging momentum, and I worried they'd find their heavy burdens difficult to pick back up should they set them down. Virgil seemed to sense this and called out to the two. "Hello! Why don't you rest with us a moment?"

The men looked back, noticing us for the first time. Their eyebrows twisted up in unison as they took in our lack of gear and preparations. Virgil explained that he was giving me a tour of the island. At this, the younger man drooped, then flopped backward into the brush along the side of the path.

"Well, ain't that lucky?" he said sarcastically. "Why are you here? What's your background?"

"I'm an engineer who leads a team of other engineers," I said defensively. "My degree is in computer engineering, but when I started getting into data science, I saw that I could use data to help my teams make smarter decisions—decisions on where we could save money, or what part of our system is the most fragile. My team is in the weeds right now, and if I can't figure a way out of it, we're all going to lose our jobs." I pointed to Virgil. "He invited me to spend some time working with him. That brought us here."

"An engineer," the man said, puffing up, "with no real data experience? Man, if that's not something else. Look. I've put in my time. Spent years working, learning the skills, making the tools. I've published three libraries and optimized two more for better performance. What have you done? Without him"—here he stabbed a vicious finger in Virgil's direction—"you've got no chance. You'll be here forever."

I faltered, having no answers or rebuttal.

"He is here at my request and invitation," Virgil interjected. "Today is his first day, and there's no shame in that. Your pack is heavy with the weight of experience. Try to remember the freedom of new discovery. For some, the burden is in learning. For others, the burden is in letting go. Safe travels, friends, and may your burdens be light." Virgil nodded to the men and led me onward.

"I don't think that guy liked me very much," I observed.

"Don't let it trouble you," Virgil responded. "There is no shortcut to wisdom. Understanding develops like tree rings. It grows over time through cycles of hardship and triumph, not all at once. Except perhaps at the beginning, which is why I hold hope for you. The choice to walk the path at all is the most important one you'll make." He looked at me and flashed an encouraging smile. "And I feel encouraged that this may work out."

We continued up the path for a mile or more. As the sandy soil gave way to a cushioned loam, I heard human sounds of hustle and bustle—a blur of voices, materials dragging over ground, and clangs and thumps of unknown origin. Cresting a rise, we came upon a grand market. Commerce was taking place chaotically at makeshift stalls that were pieced together out of tables and chairs. A barter economy had developed; here, a book was worth a spyglass, and a spare pair of boots was traded for a sturdy walking stick. Everyone buying was also selling. When a trader was happy with his inventory, he quickly vacated the stall so someone else could take his place. All around, people were barking out the things they had on offer.

Up the hill beyond the market, numerous travelers moved along the trail. Some were going up, their packs laden with fresh supplies. Others were returning with shoulders hunched in disappointment and foul expressions splashed across their faces. I watched them wander back to the market to trade for different supplies and tools, whatever they anticipated needing to reach the next plateau.

"Is the summit really that hard to reach? Why are so many travelers failing to make it up the path?"

"Listen closely to the trading going on," Virgil responded. I walked within earshot of the closest stall to eavesdrop. A tall, raven-haired woman dressed in a tan hiking shirt and green shorts had spread a hand-drawn map out on the stall's counter. At the bottom was a large X drawn in charcoal and labeled "BAZAAR." At the top was another X, this one labeled "FEAST." Between them was another X, this one marked "BEAR." Paths were sketched out around the marks, some indistinct and scratched out, some darker and more certain. None of the paths got close to the FEAST mark.

Across the counter, a thin, reedy man held a coiled rope. "Look," he said, "I just don't think it's worth it. The map's good, but it doesn't show anything east of the waterfall. That's what I need."

The woman leaned back and looked around at the other stalls in the area. "You're not the only one selling rope. You won't even find the waterfall without this. I need it all."

The man leaned forward to get a closer look at the map, but the woman was quicker and folded it back up and tucked it carefully into her pocket.

"Half?" the man offered. "Some rope is better

than none."

The woman took a second and turned a slow circle, eyeing up the wares of the nearby stalls. She didn't seem to find what she was looking for and turned back to the man. "Deal," she said.

He slid a knife out of his pocket, cut the rope in half, and passed half across the counter to her while she handed over the map. She frowned and walked away muttering. I headed back to Virgil.

"What did you see?" he asked.

"She had obviously done some exploring. She seems to know where she wants to get, and there's something in the way. A bear. But she traded her only map for rope. He seemed to need the rope too and hadn't explored much because he needed the map. So maybe he's trying to see what's out there—what problems might be in the way. But what they're going to do with a rope against a bear, I have no idea."

"Ah. There are our two friends from earlier. Let's see what they do." Virgil said.

We watched as the men took in the spectacle of the bazaar. They conferred a moment, then trudged on, turning up their noses at those hawking wares. They rebalanced their existing packs and set off uphill, leaving the market behind them.

Virgil motioned for us to sit at a table on the quiet edge of the bazaar. He fetched two small cups of coffee from an easily overlooked metal canteen, and we sat, drinking in the morning sunshine. “You mentioned problems in their way. So, let’s talk about problems,” he said. “Do you think those two know what problems they’re in for?”

“Not in the slightest,” I joked.

“Let’s talk about my problem, then. My team and I built this simulation to take people through a process that helps them internalize lessons. But this process takes each person a long time to complete. I need to find a way to keep the simulation’s lessons intact for you while speeding up its pace. Thus, the problem I’m solving most directly this morning is how to best guide you through the processes we’ve created here without diminishing the simulation’s hard-won lessons. Perhaps learning more about what brought you here will help me. Tell me more about what led you to reach out.”

I took a nervous sip of coffee as I considered my answer. Virgil was a legend in his field and one of my personal heroes, and he was putting me on the spot about what I hoped to learn from him in a day. I had no idea where to begin, so I went with that.

“My problem is that I don’t know where to begin. Your reply to my email was a complete surprise. My head hasn’t stopped spinning with questions since the minute I read it. I understand the students and the simulation, but I don’t know anything about these lessons. So, I guess we share similar challenges. You said you needed to find a way to keep the lessons intact while speeding up the pace. I need to know what the lessons are and find a way to learn them at whatever pace you take."

"That sounds like a good place to start for both of us, I think," Virgil replied.

"Something you said caught my attention. So I want to ask," I continued. "You said, 'The problem I'm solving most directly.' What's the problem you're solving less directly this morning?"

"That's a good question," Virgil replied. "Developing the experience to make good decisions with data has always taken time. People naturally learn by making mistakes and reflecting on them. The simulation speeds this process up, but it still takes time. The time that it takes to learn these lessons is problematic. I want to see if, with guidance and by learning from the mistakes of others, you can learn more quickly.

"If we can be successful today, that will be the first crumb I eat from a much bigger cake. People come to me from research, enterprise, and academia. They have projects—like yours—that have blown their budgets or didn't come in on time. Their problems largely stem from miscommunication and the inability of teams to proactively, or even retroactively, internalize the source of their issues. Things get missed. When the time comes to do the work, everyone relies on their own assumptions. The data might be skewed, which leads the analyst to create a different graph, which the executive doesn't know how to read or use. Everyone's time is wasted. I want to solve that problem. I want data-driven decisions to be reliable. I want the process to be empowering for everybody.

"Before I can tackle that, however, I first need to see if it's possible to speed up the rate at which a person can learn these lessons. If that's possible," he said with a wink, "then I'll need to figure out how to replicate your experience here. There is only one of me, after all. I can't give the whole world a guided tour."

I slurped the rest of my coffee, ruminating on what he said. He let me think it over in peace. My thoughts wandered back to my team. I thought first of a recent fight that Jan and K.T. had. I'd asked the two of them, along with their junior, Val, to look into the throughput of one of our services: the number of calls processed in an hour. The two of them had disagreed from the start about what that meant. Jan, ever the literalist, had started working on evaluating precisely what I had asked for. K.T. had put up resistance, claiming that we should really boil it down to cost instead. Her point was that the real business value would be to have the same work done for less. She'd referenced a project that she was working on with Elise that seemed to be going over budget. His response was that raw throughput could be attacked directly in the code, his code, while cost would involve a bunch of outside people and take longer to resolve. Val hadn't weighed in for one side or the other, and I had judged her for it at the time.

CHAPTER ONE

The debate had been bitter and had taken two days. I'd wanted them to work it out between themselves, so I'd let it run. That was a mistake. Eventually K.T. had agreed to Jan's side, but she didn't put much time in to help with the actual work, so what should have taken two days had taken Val and Jan a week. Looking at it now, it seemed like Val had just been waiting for the problem to be nailed down before starting work. Through the lens of my recent conversation with Virgil, Val's trepidation didn't seem like such a bad thing.

I was pulled from my own reverie when I saw the two men we'd met earlier trudging back downhill toward the marketplace. Their shoulders hung weakly, and both were covered in sweat. The old man's pack was broken, and they were half carrying, half dragging it along the path. I pointed them out to Virgil, who watched them finish their descent.

"It appears all of our friends' preparations weren't enough," he said. At this point, the two men saw us and started heading for our table. Virgil waved to them, urging them to come sit with us. I thought of extending an olive branch and ran to grab them cups of coffee and cool water. When I returned, the two had set down their packs and were catching their breath. I noticed the raven-haired cartographer had also chosen the table next to us and was surprised to see another hand-drawn map in front of her. The one she had sold had been a copy!

"Here, have some coffee and water," I said to the men, passing out the refreshments and taking a spot next to

Virgil. "What happened up there?"

"Well, up the path a ways, the trail continues to a sheer vertical cliff face," the tall one began. "When we got to the cliff, we figured we'd climb it. As we started our climb, we encountered a beehive. The bees seemed none too pleased we were there. I got stung twice, so we started to go along the cliff away from the bees. That's when the bear appeared."

"I think he was looking for the honey, and we were just in the wrong place at the wrong time," the older man chimed in. At this point, I noticed the woman at the table next to us tilting her head so she could hear better.

The short man continued, "Yeah, so we started yelling and banging things to scare off the bear. That's when one of the straps on my pack broke. And the pack is too heavy to carry without both straps. I can barely manage it when it's *not* broken. Once we scared the bear off, we still had the problem of not being able to climb while keeping two hands on the pack, so we had to head back down here to see if we could rest and regroup a bit. Thanks for the coffee, by the way. This hits the spot."

"You're welcome," I replied.

"Hey," he said, looking down and moving a small pile of dirt around with his heel. "I wanted to talk to you about earlier. I came off like an ass. What I said was wrong. To me, it felt like you were cheating somehow. Not learning the hard lessons. Maybe I'm just bullheaded enough to have to learn lessons the hard way. Otherwise, they don't feel meaningful, like I didn't have to work for it. But if you can pick up this stuff without having to fail time and again so the lesson sinks in, that's an asset. Not a liability. So, I'm sorry."

"Thanks," I said. "I feel like I'm not really meant to be here, you know? Everyone here has years of experience in data analysis and data science. I'm just an engineer trying to figure out how to make better decisions based on data. It's tough to not feel like an impostor, especially here."

"Sorry about that—I really should have kept my mouth shut," he said. "I think it's cool that you're an engineer trying to solve your problem with data. That's a rare thing. The last thing I want to do is shut you down on that." He paused. "Hey, what would you do about the climb with the bears and the bees?"

Virgil stopped him. "Your problems are your own to solve, my friend. We can't help you there. But I'd encourage you to think about narrowing your focus. Bees, bears, cliffs, equipment failures. Who knows what else is out there?"

"You do," the man replied sardonically.

Virgil laughed. "Yes, I do," he said. "But that's all the help I can give you. You'll have to excuse us; we need to be going."

We stood and said our goodbyes. Virgil and I took the coffee cups and napkins to a nearby trash can. As we left, the cartographer sidled up to the table, holding the rope in front of her. We were too far away to hear the conversation but watched from a distance as she gestured to the pans, then got to work using her rope to rebuild the broken strap.

"I think they're going to make it," Virgil said.

We left the market and climbed higher. As we walked, I started to worry about the bear. "So, what are we going to do about the climb, the bees, and the bear?"

"What indeed? And what of the bridge with the troll underneath, or the river rapids? Or any of a million other problems lurking out there? Everyone has skills, so naturally there are problems they are well equipped to solve and those they are not equipped to solve. For a skilled climber, the cliff is the easy problem. For a beekeeper, the bees are no trouble. If you try to solve a problem you are ill equipped for, it can start to feel like that one problem has become three or seven, each one an impossible act. When you bring all these problems together in your mind, you have no choice but to give up in the face of such a Herculean challenge.

"So we pick our problem and our preparation carefully. As you can see, we have little equipment." With this, he fished a small orienteering compass out of his pocket and held it face up on his palm. "I have no skill in climbing or fording rivers or talking trolls into allowing us safe passage. I am a guide. There is a hidden path that can be uncovered by means such as careful exploration, survey, or walking abreast in a line. This compass will lead me to it, and so we will follow it, and thereby choose not to solve all the other potential problems."

Virgil led us up a series of entwined switchbacks with

paths forking and diving. At one point, the path had been washed out by a mudslide. At each fork or point of question, Virgil consulted his compass and confidently chose a path. With the help of my sure-footed guide, I was able to pick my way over the debris. As we climbed higher, I could see the cliff off in the distance, rising above the trees and further beyond a cascading waterfall. We climbed higher and higher, eventually rising over a small hill. There, our path met another that led up from the cliff. Remembering the toils of the two men, we walked past the path and continued on our own way.

CHAPTER TWO

Above the rise, the trail widened and became more even. Rocky ruts turned to gravel and eventually to cut stone. We were getting closer to civilization. Soon, cobblestone streets, bunkhouses with small gardens, and storehouses appeared along our path. My guide led me onward in silence, threading through the village toward some undisclosed location.

Virgil turned a corner and beckoned me through a low doorway. The doorway was set into the side of a grand wall. We ascended a set of stairs, emerging onto a balcony above a massive banquet hall. The scene stopped me in my tracks. An enormous table of dark oak ran down the middle of the hall. It comfortably sat at least forty people. An ornately carved stone fireplace, ablaze with a large fire, dominated one end of the hall. Opposite it was a door that matched the fireplace in material, size, and detail. The hasp on the door was made of a single piece of iron which hung ten feet from the ground. It would have taken the strength of a titan to move that door.

Fortunately for the students of the institution, just such a titanic figure occupied a throne-like installation at the head of the table. Walnut-colored braided tresses framed her broad face and fell nearly to her waist. She wore a formal purple gown topped gaily with a white silk scarf. Her manner was attentive and gracious, her movements fluid and graceful—no mean feat for someone a dozen feet tall. As we entered, she was overseeing the clearing away of the remnants of a feast and the start of preparations for another.

The former group seemed to have a million details requiring her input. People asked about where to store dishes and dispose of food. Others had questions related to tending the fire and cleaning the floor. She was neither flustered nor overwhelmed. To each in turn, she gave her total focus and attention, intuiting what they needed and providing answers and guidance.

In her presence, I felt welcomed and encouraged. There was no particular quality or mannerism that made me feel this way; the whole of her being radiated warmth and acceptance. Each time her attention turned to me, my anxiety and trepidation faded, and I felt confidence and an eager, subtle swagger instead, as though anything were possible in her presence.

As the cleanup wound down, Virgil caught her eye and beckoned her over. She drew herself up to her full height and crossed to the balcony.

"Andrew, I'd like you to meet Hyatt. She is the gatekeeper here. Hyatt, this is Andrew. He is my guest, and, while he is not a traditional student, I was hopeful you might allow us passage."

"It's a pleasure to meet you, Andrew," Hyatt said warmly. She extended a massive hand for a handshake, and when I extended mine, she gripped my hand gently and shook, her fingers wrapping halfway to my elbow. "I have high expectations of those I allow through, so I must insist that you join us for a banquet. Afterward, if you can answer my questions, I might allow you to proceed."

After making the long ascent and standing to watch the preparations, the prospect of a soft chair, a warm fire, and a feast sounded tremendous. I suddenly had a pang of shame called up from my experiences in church basements long ago—I had nothing to contribute! "Thank you for your generous invitation. I'm afraid I haven't brought a dish, though. Is there any way that I can help?" I said, regretfully.

"Take a look around," she said. "It has already started."

And so it had. While we'd been talking, three dozen students had filed in through a door beneath our balcony. Some had taken charge, directing others in the placement of dishes and preparation of table settings. Others were less confident and efficient in helping with the food service. They appeared like boulders in the stream, forcing others to work around them. Curiously, no one in this group consulted Hyatt the way the previous group had done, and so she retired to her place of honor at the head of the table.

When the dinner was set, we descended the balcony and took up two empty seats. Virgil sat at the right hand of Hyatt, and my seat was on the same side nearer the middle of the table. The students took their seats one by one until the only empty seat was the one directly in front of the fireplace, opposite Hyatt. The last person standing was perhaps the least confident of those assembled. He wavered for a moment; then, gathering as much false confidence as he could muster, he addressed those assembled.

CHAPTER TWO

"Thank you all for coming to this . . . ahhh . . . banquet," he began. "I'd really like to thank our host Hyatt for the opportunity to gather here today. We will start our feast with—"

At the mention of her name, Hyatt had turned her attention to the man. Feeling her eyes on him, he froze. Gawping, he stood in place and said nothing.

As the silence and tension in the room grew, Hyatt prodded him with a generous response. "Thank you for having me," she said. "I believe you were going to describe what we will be having."

The man was knocked back into his task at hand. He took a breath and continued.

"We've spared no effort this evening," he continued, his relief and gratitude palpable. "We have five courses tonight. We start with a delectable meat and cheese charcuterie, and we will then move to a cream of potato and leek soup. This will be followed with a garden salad. For the main course, we'll have lobster roll with mushrooms. Dessert tonight will be fruit tartlet with ice cream and butterscotch. For beverages, we have a dry chardonnay with mild oak notes, a Belgian-style wheat beer, and fresh apple cider—pressed today."

The man sat, relieved to be out of the spotlight and seemingly proud to have committed the menu to memory. Everyone looked to Hyatt, who placed her napkin on her lap and chose a few bits from the charcuterie board. As she began to eat, any remaining tension in the room receded.

Conversation broke out as the students broke bread together. One of the men near me had grown up in the hinterlands of the Upper Peninsula of Michigan, a place I had spent considerable time. As we made small talk of snow machines and long summer nights, conversation flowed around the table. At one point, the table went mostly quiet, and I was able to pick up most of a conversation Virgil and Hyatt were having about the challenge of managing power and gender dynamics in a group.

With each course, the mood eased, and the conversation became more fluid. By the time the main course arrived, there was a comfortable, relaxed atmosphere. A lot of effort had gone into the production of the meal. Each dish was well made and came out on time. I enjoyed the meal, thinking the students should be proud of their efforts. As I looked around the table, I saw contented smiles on most faces, though a few people had leaned heavily into the salad without touching much else. They didn't look happy, but more as though they were trying to keep up appearances for the sake of the group. The man who introduced the menu also hadn't eaten much, which I chalked up to his nerves. The face most soured by the meal was, to my surprise, the host's.

When everyone had eaten their fill, Hyatt shifted her massive chair back from the table. She picked up a wine glass and rang it with a fork to gather the attention of anyone who hadn't already noticed the twelve-foot-tall titan gathering herself for a speech. The room fell silent.

"First, I'd like to thank you all for your efforts. The meal was most delicious, lovingly and skillfully prepared, nutritious, and well executed. You have my gratitude for all the effort that went into its preparation. With that said, I have some questions. Who was it who planned the menu?"

The meek man at the far end of the table hesitated, then responded. "We all did, we . . . we did it together."

"Wonderful," Hyatt responded dryly, her face betraying complete disagreement with her words. "How did you divide up the work?"

The woman sitting on my left responded. "We started by figuring out what people needed. Some of us are vegetarian, some are vegan, and we have one who is gluten-free. We split into groups for each dish and made sure there were people with different needs assigned to each course. Then we broke off to plan and cook.

"For example, I was on the charcuterie team. We had the gluten-free person and some vegetarians. We thought some bite-sized finger food would be a good way to start the meal, so we got some different cheese and meats. The dish was simple, sure, but it was tasty, and we spent a lot of time pairing the meats with the cheeses and spreads."

Hyatt smiled. "Did you think to add some crackers or snackable vegetables? Maybe some pickles, celery, carrots, almonds, or hummus? You had to know that there were people who wouldn't be able to have the meats, cheese, and spreads."

"Well, we knew there were a lot of vegetables going into the soup, so we thought that we'd stick more to the finger-food protein options."

"All right," said Hyatt. "How about that soup. Who worked on it?"

A man spoke up to her left. "I was on that team. We were initially thinking of a straightforward potato and leek soup recipe. We got started on that, but then started talking with the salad team about what they were planning. When they went simple, that left us the chance to go a bit richer, so we switched to a cream base to take advantage of that."

Hyatt nodded. "Did you talk with the first-course team about what they were planning?"

The man paused. "No," he admitted. "They asked us what our plan was and we told them, but we didn't think to ask the same of them."

"What about the salad? That was a well-balanced dish," Hyatt continued.

A woman sitting next to Virgil spoke up. "I was on the salad team. We had two vegans and two who like meat. We came to a quick agreement that something light and nutritious would be a nice contrast between the rich soup and the main course. Garden salad was a great choice, but the meat eaters wanted to include chicken or ham for extra flavor. We compromised on veggie-bacon-flavored bits, which are vegan but still add texture and flavor to the dish. I think we did a great job."

"The salad was tasty and a good choice," responded Hyatt. "I appreciate the thought and effort that went into it. Would any other teams care to walk me through their choices?"

A man near the fireplace stood up. "I worked on the main course. We had mostly people who eat meat, although I only eat seafood and we had one vegetarian. We wanted something lighter than red meat, due to the heavy soup. With my love of seafood, I convinced them that a lobster roll would be just the thing. Our vegetarian then decided on the mushrooms. I thought that added a lot of depth to the dish. It was a good addition."

"It was a balanced dish that was well prepared, that much is true," said Hyatt. "But once again, you seem to have left a bunch of people behind. Anyone who can't eat gluten can't eat the crust and therefore might not touch it at all, not to mention the vegetarians and vegans. A side of mushrooms, no matter how well prepared, is a weak main course. Dessert folks? How about you?"

A woman near the fireplace stood. "The dessert team waited for the rest of the menu to be decided before we acted. Really, we started with the drinks. With that rich food, we wanted wine on the drier side and a beer to match. We thought the sweet fruit of the tartlet would offset the dry of the wine really well, and figured the ice cream could be added by folks who would like that. Otherwise, they can leave it off."

"Is there dairy in the crust of the tartlet?"

"A bit, yes. We needed to layer in cold butter to make it flaky."

"I see," said Hyatt. She cleared her throat gently and addressed the whole table. "You look at this table and see success. You see what you contributed and take comfort in your small part of the overall plan. However, you fail to see the bigger plan, the composition of the meal as a whole. If you did, you would have made vastly different choices. What looks to you as success I see as utter failure. Myopic. Narcissistic. Failure. You have, from the moment you forced the one among you with the worst case of stage fright to introduce you, failed to identify and account for the needs of your peers. You knew some among you don't eat dairy or meat, and you handled this knowledge by leaving them out of the first two courses, barely accommodating them with the main dish, and abandoning them again for dessert. They got scraps and salad. *Scraps and salad.*"

The vegans sat back as they were mentioned, smug smiles on their faces. One gave a shrug as if to say, "What can you do?" It seemed being left out was not a great surprise to them, and I suspected the experience was part of a pattern in their lives, one that was probably revisited every time they spent a holiday at a distant relative's home. They seemed to have resigned themselves to the margins.

Hyatt continued her critique. "I find it telling that some

of you are now sitting back, grinning and playing the victim in this. You participated, saw the direction things were headed, and never stuck up for what your actual needs were. You are as complicit as the person who failed to mention he was lactose intolerant," she added pointedly, staring down the table.

The meek man near the fireplace shrunk even further and stammered out an explanation. "I—I'm sorry, I just don't like making things more complicated," he moused. "Besides, I can usually do all right by picking around what I can't eat."

The titan waited for him to finish, then continued. "That brings me to the composition of your menu. You subdivided and delegated the work, but then nobody thought to look at the whole menu. Every dish but the salad left significant numbers of people out, but you thought since they had been included elsewhere that their needs were met. You were to cook a meal! A feast! What you brought was a jumbled mess of exclusionary dishes that left almost everyone wanting at some point."

Her voice gained volume and tempo as she came to her conclusion.

"You failed to actively discover your needs! You failed to advocate for your needs when they were neglected! And you failed to make those needs fundamental to what you made! Across the board, you have failed."

Hyatt slammed her fist on the table. The force of the

blow should have sent glasses tinkling and cutlery clunking, but the entire hall was silent. Everything involved in the dinner service had just been turned to dust.

Plates, cups, knives, spare pieces of lobster roll, and drabs of salad dressing—all was dust. My mouth went dry, and gnawing pangs of hunger, unexpected after concluding such a feast, grew within me. It was clear others felt the same.

Hyatt dismissed the students without another word. They swept up the dust of their failure and departed through the side doors without speaking.

Virgil and I accompanied Hyatt into an antechamber, where she was to make good on her earlier promise. "Thank you for dining with us," she said, "and for enduring the less compassionate part of the lesson."

"Thank you for having us," I responded mechanically. My body was still processing how to feel about having a meal taken away in such a manner, and I could muster none of the wit I might have employed on a better day.

"What do you think they did right?" she asked.

"I thought the food was excellent." I said, defending the group. "Each dish was delicious, and together it was great from start to finish."

"This is because they built it, coincidentally, to suit your needs. You get partial credit. Someone is always happy, and so the group can be fooled into thinking everyone is content even though each of them is, in fact, starving in a way they're willing to hide." Hyatt paused a moment. "Such a waste," she muttered to herself. "If you don't have a grasp on this concept, I hesitate to allow you passage. How about the other side—what do you think they did wrong?"

"It seems to me the thing they missed was the chance to appreciate individual needs together, as a collective, *before* separating into groups. Maybe if they had done that, they could have unified their vision and advocated across courses for all the needs they'd heard called out, even if those needs weren't personal. That approach might have encouraged people to speak up. It can be easier to stay quiet if you feel like making your needs known slows the group down from making progress on a big task."

"I disagree," said Hyatt. "They did that, to some degree. They would have gotten it even more wrong if they hadn't discussed needs among their smaller groups."

"Oh!" I interjected. "They didn't come back to those needs after finalizing the menu. If they'd had a quick discussion at that point, then people would have had a chance to speak up, either for themselves or for others they noticed being left out."

"Half credit on that too. You're right, they need to keep everyone's needs in mind throughout the process, not just at the beginning. Unfortunately, partial credit doesn't get you through," she finished.

Virgil, who had been quietly leaning on the doorframe throughout our discussion, stood up and addressed Hyatt.

"Could we arrange some sort of extra credit?" he inquired.

She thought a moment and responded, "If he could tell me about a situation that he is screwing up in this same way—and how, having been through here, he'll fix it when he leaves—that would do it. But you must also promise to send me a note telling me whether that fix worked," she said, turning her attention to me, "and the example you give had better be good."

"That seems fair," Virgil replied.

"I'll need a minute to think." I said, scrambling. The project with Jan and K.T. was top of mind, but it didn't seem to fit initially. K.T. and Jan had a good understanding of their users, but something about the project wouldn't let my brain dismiss it entirely. K.T. had been focusing on price, not out of business interest, but because that had intersected with the other work she was doing. That mixed in my head with a story from way back, which exposed the true issue at stake. I began to explain.

"I need to put the current issue in context for you, so let me start at the beginning. Seven or eight years ago, I was working on a system that provided pricing information to various other parts of the business. One year, when doing our yearly updates on the service, I got a panicked call from a department head in a different division that I had never heard from before. 'Did you change the pricing service recently?' he asked, to which I replied, 'Yes, we just did our yearly changeover and some of the data fields were updated. Why do you ask?' Long story short, his department had integrated without telling us and had built a sales-forecasting system that now fed our entire sales team. And now they were down because they weren't in the loop for our changes. We had users that we weren't aware of, so we couldn't anticipate their needs."

Hyatt nodded, asking, "So how does this relate to a problem you can fix with this lesson? Surely that issue has long since been fixed."

I inhaled deeply, knowing how complicated this was going to sound. "I think the same problem is repeating itself on my team right now. On the project my team is currently building, one of my senior engineers paired with one of my junior engineers, and they got it built quickly. So quickly, in fact, that the main users of our service—the marketing department—haven't been able to adopt it fully. I feel like their needs have been well considered, and they've been great partners in planning and building the service.

"The weird thing is that we've seen usage of the new service take off, wildly exceeding our expected usage considering that the marketers haven't been in there much. We saw so much usage that it blew up our budgets and the senior engineer started looking to make up for the shortfall in other projects, which put her at loggerheads with another senior engineer on the team. At the time, I figured it was just two strong personalities with different perspectives, but that's not all of it. If marketing isn't in there, who are those users?

What do they need? K.T. might know; she's been around long enough that the current users might have gotten their information through her. But I've got to get ahead of this. Otherwise, it's going to be the same thing all over again."

Hyatt took in my story and thought a moment. "Tell me more about how you'll prevent this from happening again. Not knowing your users is worse than not knowing their needs."

"I know, I know," I stammered. "In this case, I think we can set up some monitoring for where the incoming calls are coming from. We know what marketing uses to call in, which means anything off that list should be our mystery users. Maybe we can even bill them back for our extra usage. That might ease the budget issue in the short term."

She looked up at Virgil. He shrugged. "It's up to you," he said.

Looking back at me, she said, "If you can pull all of that together, and you promise to let me know if it works or if it fails and how, I'll let you pass."

A wave of relief washed over my whole body, as I realized I had tensed up during the questioning. "Thank you. So much," I said.

She led Virgil and me back to the banquet room and flicked open the massive hasp that held the door shut, its weight yielding to her as a page yields to a reader, like an afterthought. The door opened to allow us through. Virgil led the way. As I passed her, Hyatt pressed a small parcel into my hand. "For the journey. You'll need the strength." She smiled warmly and closed the door behind us.

The package was warm to the touch. I gave it a quick sniff. The smell of warm lobster roll and leek soup filled my nose. Ever the proper host, she had seen to my needs.

CHAPTER THREE

Through a short tunnel, we emerged from the hall into a blinding sunlit basin. A path which climbed clockwise from the bottom tracked a steady course out of the basin. The path was close to the edge, and each wayward step sent pebbles cascading toward the center. Around the ring, we could see dusty clouds kicked up by the feet of many groups plodding their way toward the top. One of these groups passed us. We hurried to catch up.

"How's it going?" asked the group's leader, a thin man of maybe forty.

"We are well," Virgil responded. "May we travel with you for a bit?"

"Sure! You can walk with us until we reach the castle," he offered, gesturing ahead toward a distant stone structure that peeking out from surrounding cliffs. "But you'll need to go in by yourself. We've had a hard enough time keeping our stories straight even without two more people mucking it up."

"What stories do you have to keep straight?" I asked, confused.

"When you get to the top, to get into the castle, you must get by a gatekeeper. We got up there last time, and he asked each of us what our goal was and how we were going to achieve it. I gave my honest answer, and I think everyone else did too, and so he opened a door for us that we thought would take us inside. Instead, it put us on a path back down into the center of the basin. We've been hiking for three days to get back up there," he explained.

"That must have given you plenty of time to talk through your answer," I replied, trying to be optimistic.

He looked from me to Virgil and then stopped, his face screwed up, thinking. "Wait, are you Virgil?"

Virgil sighed, nodding. "Yes, I am Virgil. I am here to guide Andrew through the journey."

The man beamed. "I remember you from our orientation! You designed the machines, the whole world here. It's great to see you again! Hey, I bet you can help us. You've got to know what we need to do! What does the gatekeeper want?"

"You've had three days to think about what you each said. It sounds like you feel your time has been wasted, and that you are frustrated. Tell me about that," Virgil deflected.

"Time is a weird thing here," the man replied. "I'm frustrated, but not from impatience. We have all the time in the world. I guess the frustration is with my—our—delay in figuring out the lesson. I get angry at my own difficulty learning. I've always heard you only really learn after reflecting on what you've done. So, the three-day walk is a good thing, kinda. If we just stood at the gate trying things until we got in, the lesson would be just trying things until something works. I just don't think that's it in this case. I just wish I knew what *it* was."

I was intrigued by his stoicism, but the castle was still miles away and not getting any closer. "Can we walk and talk?" I asked. "I'm sure you're as eager as I am to get up there."

We established a steady uphill pace, and Virgil begged him go on. "So, what do you think the lesson might be here?"

He thought for a few moments, feet crunching against the dusty gravel. "Well, I've been trying to figure that out. There are a couple of parts that must work together. You've got his two-part question, 'What is your goal, and how will you achieve it?' I've solved problems before; that sounds like he's asking for what my solution is. A goal and a method. It's got to be.

"He asked us as a group for our solutions. Something was off about the way we answered, clearly. So, we were sent on a long path to arrive back at the same question. There's a not-too-subtle message of time wasting there. So, when you put those together, it seems like something about our solutions caused us to waste our time. That seems clear. I just don't know what it was."

Virgil spoke up. "If I were to tell you the lesson, you wouldn't properly learn it. You nailed it on the head when you said that reflection will lead you to true understanding. It will take some time to internalize the problems that lead to the lessons and *why* they are so important. I won't take that away from you. You need to do this on your own."

We walked on in silence until we had nearly reached the castle. Then we let the group go ahead of us. I stopped with Virgil and took a rest to eat and think. I unwrapped the lobster roll and ate my fill. After the lobster roll was crumbs and memories, Virgil spoke. "What did you think about what he had to say?"

"Well, I couldn't disagree with his conclusions," I began. "But I think he missed another factor. If it were just a question of whether the solution he offered was good or bad, there would be no reason for the group to be evaluated as a whole. But they were, so there's got to be some kind of group aspect to it. Until he understands that, I would bet he'll keep going around until he has a better understanding of what the gatekeeper wants."

"So, what does the gatekeeper want?"

"How should I know?" I sputtered. "It doesn't sound like he gave them a specific problem to solve. For any of the solutions, a million fine details could be at play. Any one of them could be used to disqualify the group. Honestly, it doesn't seem fair to me."

"You're right."

"I am? About what?"

"If the gatekeeper were to judge based on the details of their proposed solution, it would be entirely unfair. It would also be futile and a waste of time. They'd be here forever!" Virgil said with a chuckle.

I thought back to the previous lessons. At the bazaar, Virgil had expressed hope for the three travelers only after they started working together. Then, at the feast, all the cohort needed to be included. Could that be it? I wondered.

"It has to be related to the group, then," I said. "That's what they missed. They're not here each on their own, they're a group with a problem and a solution."

"Have you ever had a team that would agree to a problem but come up with different solutions?" he asked.

"Yes."

"Did you end up implementing all the solutions?" he continued.

"No, that wouldn't make any sense. You pick one. Oh," I said, finally getting the point.

"So that's it? They just have to say the same thing?" I asked incredulously.

"If they don't even start with the same description of a solution, what do you think will happen when they go to build it?"

"Well, I guess different groups might build things that don't fit together. And that would waste a bunch of time and money," I responded, grasping the point.

"There you go. Now, we need to get through the gatekeeper as well," Virgil said. "Unless you want to go for a full tour of the basin here."

"Not particularly."

"Then what would you say is our goal?" he asked.

I wracked my brain, trying to phrase our common ground simply. "We're both under a time crunch, fighting to discover what's possible. We both want to help people with information we don't have yet. We've got to learn first."

Virgil gave me a smirk. “So, learn, grow, and change the world. Sounds simple enough.”

“Learn, grow, and change the world.” I repeated. “Learn, grow, and change the world. Can we write that down? Something about physically writing something plants it more firmly in my head.”

Virgil glanced around and snatched up a stick from the side of the path. With it, he scratched “Learn” into the dust of the path and handed the stick to me. I looped through the letters in “Grow” and handed the stick back. He finished the message, and we both stared at it a second.

“Are we good?” asked Virgil.

“One more thing,” I asked. “Couldn’t they just write it down too?”

“If they ever want to get past the gatekeeper, I’d hope so. Have you ever tried to get ten people to say the same thing from memory? Impossible, except if you put it to song, maybe.”

With food in my belly, we covered the last short walk to the castle, armed with Virgil’s plan. The group who had walked and talked with me had already gone through. Whether they made it forward or were doomed to repeat the basin, I did not know.

We presented ourselves to the gatekeeper, a short, stocky fellow with two large keys on his belt. He seemed to recognize Virgil but still maintained his station. “What is your goal, and how are you to achieve it?”

I thought back to the words scratched on the path and confidently answered, "Learn, grow, and change the world."

He asked the same of Virgil. "Learn, grow, and change the world," was his response.

"Your goals and solutions are in harmony with each other's, so you may pass." He unlocked the main gate, revealing a wide, paved avenue heading uphill toward a large park. Virgil thanked the gatekeeper as he shut the gate behind us.

CHAPTER FOUR

With my guide, I entered the park and sat beneath the arching branches of an old maple tree. "I need to rest a moment—is that all right?" I asked.

"Sure," Virgil replied. "We have made good time and have plenty to spare."

"Can I ask you something?"

"Yes, of course. What do you want to know?"

"Why did you open your doors and your process to me?"

"Our process has been a trade secret for a long time. The graduates are successful, and our program is self-sustaining. I wanted to see what could be learned by a novice in a shorter amount of time.

"The lessons learned here, and the time spent agonizing over them, lead to a valuable economy of effort in the future. Graduates run targeted projects with very few wasted resources and a high probability of success. These projects are planned and executed for maximum change in minimum time.

"However, the institute has a limited capacity—we can only take so many students at any given time. I want our mission to expand beyond what the walls of our institution can contain. The basic lessons acquired here can empower and encourage people far beyond our walls. The simple ability to acquire and use data to power decision-making is, on its own, far more powerful than any, perhaps *all*, the other steps combined. It moves you from making decisions based on your *perceptions* of how the world is, to making decisions based on how it *actually* is.

"Sure, you can have bad data or misleading analysis, but here's the big secret—when someone discovers a way to improve their data or fix a flaw in their analysis, they will naturally reconsider the conclusions they made from that bad data or analysis. Something has fundamentally changed, they realize, and it's time to reconsider. You don't get that with gut instinct. It's far harder to get someone to reconsider a conclusion they've made if they haven't based it on data and facts, but on their gut. Those unmovable conclusions can be costly and dangerous. If we could get this message out more broadly, maybe we could do some good."

I felt compelled to know whether Virgil's time with me seemed to him to be well spent. "How am I doing as a student so far?"

"Better than I thought you might. You seem to be grasping the core concepts, despite the limited time we have together. I was afraid it might all go over your head. It hasn't. You've done well so far, but I guess we'll see how you do in the quarry."

"Does that have any implications for the school? If I can do this, are you thinking more about rolling out a quicker cycle?"

"Yes, it does. If we were to change the format, there would be massive upheaval to the institution. We'd need a new generation of units built for processing power instead of longevity. Many of the staff, including Hyatt and the gatekeeper, would likely be let go. We would draw our student population from different circles, and we'd have to figure out what to do with our existing students. Then there's the question of *my* role in all this. There's only one of me to go around. Making digital copies of myself at a fidelity that's able to instruct and guide is quite the mission! Exciting, but it leaves me with lots to think about. Without real students to learn from, we'd have to provide virtual students, staff . . ." Virgil trailed off, deep in thought.

I let him ponder a moment before asking a question I thought might get us moving again.

"You mentioned something about a quarry?"

"Ah, yes. Thank you. It's right this way," Virgil responded, his mind still clearly immersed in another thought.

Virgil then led me on a long path that sloped upward through the park. In time, we came upon a small hill that allowed us a view of a wide, sunken pit abutting a sheer cliff face. I could see nothing above the cliff face and thought that perhaps this might be the summit. We were now closer than ever, though how we would scale this cliff I could not guess.

The quarry itself stretched far off into the distance, wrapping around the peak. Small holes pockmarked the quarry, as though an army of prairie dogs had been hard at work digging burrows. Dark recesses in the cliff walls were flanked by high piles of discarded stone. Virgil saw me staring and explained, "Here is where we collect materials.

Some students mine their own, but you can also barter or buy what you need."

"Materials for what?"

"For their great project. You'll have to see it; an explanation won't do you much good."

I followed curiously down into the quarry. All kinds of things were available. There were plenty of publicly available tools for those interested in mining themselves, and there were also piles of materials that were free for the taking: piles of iron and copper ore, uncut gemstones, and a rainbow's worth of geodes. People were offering materials for sale or trade alongside specialized tools to work them.

I approached one of the vendors, fascinated by the sheer variety of his goods. "What are you looking for?"

"What do you have?"

"I work on lenses, mostly. Glass lenses, quartz lenses, even some ruby lenses if that interests you."

"I'll think about it," I said, playing coy about my utter lack of anything to trade. "But why should I buy from you?"

"Great question! Because I'm the best, ask anyone." The vendor smirked, an expression that faded when Virgil piped up.

"The best is kind of a stretch, wouldn't you say?"

"Well, OK. Maybe not the best. But I'm good. Sure, you could grab some free glass from over there and grind out a lens yourself. But do you really want to go to all that effort? Do you even know how to pick good glass from bad or how to choose a cutting fluid? Not to mention how to get a consistent profile in your grind . . . I do. Maybe you could learn all that, or maybe you could just buy it from me."

"I like your pitch, but just so you know, I don't have any way to pay you," I said. "I do like that red lens, though," I added, pointing to a crimson disk of glass with a million tiny facets.

"That's a nice one." He picked up the lens and turned it over in his hands. "I'll tell you what. I've got a few of these—why don't you just take this one. It's on me."

"Thanks, but is there really nothing you want for it?"

"Just pay it forward," he said simply. "The next time someone needs something that you can provide, give it to them. Then you can consider me paid back."

"Wow, all right, then. Thanks!" I took the lens.

"Good luck." With that, the vendor saw another group approaching his stall and left off with us to go attend to them. Virgil and I walked passed other sellers' tables to a winding path that took us around the back side of the summit. We made our way to a massive black pile that towered over a collection of small, open-walled huts, each with a trail of smoke curling out from the roof.

"What's the black pile?" I asked.

"Coal. For the forges. Ore refining, just like data refining, takes a lot of energy."

CHAPTER FIVE

As we worked our way closer to the coal pile, the air developed a strong note of hot metal, an odor I tasted on the roof of my mouth. Tucked behind the coal pile was a warren of small workshops tended by an army of students. The sun had fallen in the sky as we made our way around the quarry, rays filling the shops with late-afternoon light.

Students worked on all manner of mechanical trades. Blacksmiths refined ore into ingots, which were then hammered into sheets or pulled into wire. Gem cutters hunched over tiny wheels, carefully cutting facets and crafting jewelry settings. Wood-carvers broke down massive trunks, toiled over lathes, and manned saws. I even spied a chemist who was compounding chemicals, for what arcane purposes I did not know. The focus and hustle of the students were palpable and delightful. I spent the better part of an hour surveying their different crafts.

I found one artisan more captivating than the rest. When I first saw her, she had a mountain of steel rings in front of her. In each hand she gripped a blunt, smooth-jawed pair of pliers. One by one, she gently bent each ring into a helix. Over and over she did this, lining up the helixes into rows and columns, packing them tightly on her wooden table.

Once she opened every ring in the pile, she dug through her bag and pulled out a pouch filled with leaf-shaped steel plates that had been polished to a mirror shine. A hole the same size as the rings had been punched at the base of each plate. With careful precision, she began to thread the rings and plates together to form a chain.

When the ring-plate-ring-plate chain was as long as her arm, she put the pliers down, stretched out her arms, and cracked her knuckles. Her hands moved methodically, rarely putting down the pliers and using all ten fingers to grab, transfer, and set the materials. Though I had no idea what she was making, I was rapt. I found the process mesmerizing.

Once the second chain was complete, she laid both chains out in parallel and started weaving them together. She made a third and then a fourth chain and bound them all together into a flat sheet of scales that reminded me of snakeskin. Then, working slowly, she bound the ends of her creation together to form a cylinder about the size of a dinner plate. With this done, she further refined and cinched together the top of the cylinder. When she was done with the final layer, she had formed a perfect shiny dome.

She turned it this way and that to inspect it for flaws. Then, looking very pleased with herself, she tucked the finished dome into her bag, stowed her tools, and rubbed her aching hands. She stood up, gave a small wave, and picked up the path we had walked in on. She walked down the path, away from the coal pile and further around the mountain. The sunset was now starting to wane, brilliant orange fading slowly to purple.

Watching her walk into the sunset broke me from my reverie. Though it felt like moments, I had watched her work for hours. It was rapidly getting dark. I really wanted to uncover the purpose of what she'd been building. I turned to Virgil and asked if we could follow her to see the next steps. He agreed, and we took off after the artisan to see what she would do next.

When we caught up with her, she was sitting at a low, wide table, emptying the contents of her bag. First came the shiny dome; then she pulled out an assembly that looked like the top of a tiki torch. She had a bottle of something, a lid, and a wick. She used some wire to hang her wicked bottle under the dome like the basket of a tiny hot-air balloon. The sunset glimmered off the dome, split into a million tiny reflections.

"It's a lamp!" I exclaimed to Virgil. "It's so beautiful."

"It is quite beautiful," he agreed. "I thought we might have spent a bit too much time watching her make it, but perhaps that is all right." He said.

"So, is that what everyone is making? Lanterns?"

"Not specifically lanterns, no," my guide replied. "But the way forward is a dark one. Some sort of illumination is necessary. The students come up with all kinds of solutions. Some are more effective than others, of course, and each requires that student's particular skills. If you'll come with me, we can leave her to her preparations. We'll go ahead a bit, so that you can see a few more projects."

CHAPTER SIX

We walked through the jumble of workshops in the growing darkness. I was afraid of stubbing a toe on an unseen table leg, but Virgil's quick, sure guidance got us through. He beckoned me forward into a clearing lined with benches. A campfire burned at the center. Students were scattered around the edges. Each one was fiddling with final details for their illumination device.

Virgil took a step toward the fire and cleared his throat. Everyone in the clearing gave him their attention. "Excuse me, friends. Tonight, we have a bit of a special occasion. We have a guest with us for the first time. Someone who isn't a student but is here as my personal guest. He hasn't completed the trials you have but is still an eager learner. If you would be so kind as to say a few words to introduce your illumination projects, it would mean the world to me."

One of the students edged over and introduced himself. "Hi, Virgil, this is my new idea. I've tried a few different approaches, but they've all failed. So, I figured I'd just go old school and make a torch. Wood, cloth, oil. It is, *by far*," he emphasized, "less complex than my other ideas. But I think that may help here." He brandished what looked like a club.

Virgil looked at it and rendered judgment. "There can be strength in simplicity. Thank you for sharing."

He sat down, and another student stood up. "Hi, Virgil! This is my first project here. These other folks probably have much better stuff, but I've been working on this twin candelabra." The man held up a silver candelabra with two pillar candles. The work was rough but consistent, and it looked quite sturdy.

"Where did you get the candles?" I asked.

"I made them. I spun the wick myself and bought some beeswax from a vendor at the market."

"That's really cool. I like them." I replied.

"Thanks. We'll see if they work, I guess," he said with a chuckle.

He sat down, and the woman I'd spent all day watching stood up. "Hi. Umm . . . I guess this is my second project. I built a lantern for my first project, and I made it out of steel and oil. It was super ugly. It took me a long time to design and build, but when I lit it on the path, the light went all around, but not down at my feet where I needed it. So that's what I set out to fix. I thought a reflective shade would redirect the light down toward the path. So, I went back and stamped and polished all the plates, drew out the wire, and made the rings. I think you saw the rest. I love how it's come out. When I designed it, I hoped it would be something between a disco ball and a chandelier."

"It's amazing. It looks stunning, and to know the function behind the form is great. Thanks for walking me through it," I told her.

Virgil spoke up. "Thank you for all the hard work you've put in here. I believe it's time to proceed. Which of you will go first?"

The torchbearer stood up, walked over to the fire, and

gently lit his torch. He studied it for a moment, and when he was satisfied with how it was burning, he held it aloft and began picking his way up the path. Almost immediately, we lost sight of him, though we could measure his progress by the light the torch threw off. Up and up it went, eventually cresting the cliff and disappearing from view.

When the torch could no longer be seen, the student with the candelabra followed suit. Gently lighting both candles, he gave an awkward wave and headed off in the same direction as the torchbearer. He faded from view. A moment later we heard a thump in the dark, followed by a cry of pain. "I can't see anything! Argh!"

The woman with the lantern called out, "You'll be OK. Can you still see the fire? Come back!"

A few minutes later, the man limped back into view with a torn pant leg. He held the twisted remnants of his creation. He slumped back onto the nearest bench and began examining the candelabra. It was a total loss. The base had broken off in the tumble. Both candles were missing, and one arm was bent at a severe angle.

"It's OK," the woman said. "It's a hard thing. Rest for a bit, then make a new plan. You'll get there. We all will, just not tonight." She rose and touched him on the shoulder to reassure him. He gave a grim smile but seemed to gain a bit of resolve, as though he were dispensing with some of the desperation that he'd held a moment before.

"Here goes nothing," the lantern maker said. She took

a small twig, lit it in the fire, and used it to light her lantern. She took a step away from the fire, and the lantern reached its luminous peak. "Wow, that's bright," she exclaimed, smiling broadly, clearly very pleased with herself.

And it was bright. The lantern light was reflected and amplified by countless surfaces and edges. As she walked away from the light of the fire, the dancing disco-ball reflection caught me in the eye, and I reflexively put my hand up to shield myself from its blinding light. The same thing happened to the lantern maker a moment later, and she wound up holding the lantern in one hand and shielding her eyes with the other. This made her progress slow at first, but in time she was able to pick her way up the slope. The remaining student excused himself and headed toward the backsides of the forges. This left Virgil and me sitting around the slowly dying coals of the campfire. Staring into the flickers of blue and red, I thought back to an issue I'd had recently when K.T and Elise ran a project over budget.

I had put K.T. and Elise on a project together with the idea that K.T. could mentor our newest team member and maybe discover some fresh concepts for herself along the way. What happened instead was that my aloof and overworked senior engineer had let the young upstart run the project. She would advise when asked, but she'd felt no need to rein in the scope of the project.

I glanced at Virgil over the glowing coals. He'd been watching me but seemed not to want to interrupt my reflection. “Is there something on your mind?” he asked.

“I was just thinking about an issue I had with part of my team,” I said. “I tasked them with evaluating a couple systems. They were to come back with numbers that we could use to pick one system over the other. I asked one of my senior engineers to lead it and tasked one of my junior engineers, Elise, to help. It became the kind of thing where I’d ask for an update and get told there was great progress, but it was still a week out.

“I didn’t keep them accountable. Looking back, it’s so obvious. Elise’s success on the team has been in automation. She’s fighting for the respect of being considered a peer among the other team members. She went and created dashboards with chord charts and all sorts of slick data visualization when all I had wanted was a quick analysis of a couple factors so we could choose one system over another.”

“So, you wanted the torch,” Virgil said.

“And she built the lantern,” I said.

“They’ll both get you there,” Virgil continued. “Both your Elise and our lantern maker looked smart and talented as the designers. Their solutions frequently dazzle the eye and distract the mind, but they need to learn that the path is what’s important, not the light.”

“So, what’s the lesson?” I asked.

“You tell me. What did you take from it?”

“Well, if I’m going to start at the beginning, the problem is the darkness. You can’t see the path. The person affected is the student; they must travel the path. The solution would then be to illuminate the path somehow."

"Easy enough. I think you're starting to see the process at work here."

"I think the task of building the light is a metaphor for problem-solving. In this place, your students use raw materials such as wood, string, beeswax, and steel. Outside of the simulation, their materials are data and analyses. In both places, they need to understand the material, shape it, refine it, and know what purpose it will be put to. That way you can ensure that your light fits the path, so to speak."

"And therefore, that all the work that went into defining and creating the solution, from describing the problem through to refining the data and settling on analyses, will efficiently accomplish what needs to be done. It's easy to get lost along the way, which is why I brought this."

Virgil trailed off and reached into his pocket to pull out a flashlight. He switched it on, and I reflexively shielded my eyes again. "Do you still have that lens?" he asked, reaching out his hand.

"Sure do." I fished it out of my pocket. Virgil took it and pressed the red lens to the front of the flashlight. The beam shone red, the light dancing a bit from all the reflected facets. He shone the beam over the ground, illuminating some boulders and small scrub trees that had been invisible a moment before.

Together, we picked our way around gnarled roots

and over boulders. We stuck to the path, switchback after switchback, guided by the soft red light. At one point, I looked over the edge and saw nothing but a black abyss all the way down to the ocean. After that sobering sight, I stayed well inside the beam of Virgil's light. We toiled up another few switchbacks to emerge onto a vast plateau. The first blue rays of sunrise were just starting to streak above the horizon. The plateau was windswept and empty; there were no students and nowhere further to climb.

"So, is this it, then?" I asked.

Virgil smiled as he answered. "No, not by a long shot. Try to relax—this next bit can be a bit jarring."

A feeling of vertigo swept up my spine and knocked on my brain stem. I couldn't tell which way was up. I was able to move but didn't really want to. I felt like I was staring through some other person's eyes. The world went plastic, as though seen through a television. My mind disconnected, consciousness aloft on vaporous wings.

CHAPTER SEVEN

I heard the ethereal music before I realized I was no longer on the plateau but in a comfortable bed. Next to the bed was a side table with a bowl of fruit and a small sign that read, "Come out when you're ready." After taking a moment to get my bearings, I grabbed an apple and went to find Virgil. The door to my room opened into a hallway with a gathering space at one end and tall double doors at the other. Virgil's voice was carrying down the hall, addressing a small cadre of his students.

"It's good to see you all again. You've been through quite a lot since last we spoke—trials, tribulations, and no shortage of hard lessons. I told you at the beginning that it was not an easy road that you had chosen, and I believe now you understand how much I meant it. I commend you for your commitment and resilience. Give yourselves a round of applause." Virgil paused a moment as a loud cheer and applause flowed through the room.

"I have good news and bad news for you. First, the good news. You are no longer students. You have completed your trip through data purgatory. But here's the bad news. You are not yet graduates. There is one more lesson you must take to heart. Each of your journeys is unique, but none of you came here equipped only with the desire to learn. That was a by-product. You came here to make change in the world. To stir things up a little. To break down barriers. Knowledge, alone, isn't enough. You must understand how to use your skills to enact change.

"When you come upon adversity and your first instinct is to start forming a solution, remember the lessons you learned on the beach—if you do not deliberately choose one problem to solve, and take time to understand that problem, you will be set back time and again.

"Once you have that one problem in hand, you will be tempted to start designing solutions. In those moments, remember the lessons of the banquet. Problems affect people. People have needs. The solution must fit the needs of all, not only the most obvious needs or the needs of the most vocal.

"When you are tempted, out of haste or sloth, to let the group splinter and start their own solutions, remember the lesson of the quarry. Agree to a solution that keeps in mind the specific problem. Then move ahead together.

"When you are tempted to use any old data that seems to fit, remember the time you spent in the quarry. The tradeoff between finding your own data and paying others is yours to make, but don't make that decision lightly.

"When you are tempted to trust data implicitly, remember the time you spent at the forges. The quality of your data depends entirely on the time you're willing to spend refining and improving it.

"When you are faced with the challenge of bringing all that hard work together, remember the lessons you learned around the fire. You can emphasize function or form, just don't leave yourself in the dark.

"And lastly, when you come to a place where you have implemented your solution, you might be left with the question *What next?* Reflect on this now. What does success mean to you? You have triumphed over an incredible journey. What is now possible for you that wasn't before? What is your next step? Write it down, share it with me if you want, and then get to it."

Virgil adjusted his posture and took a lighter tone. "To help you with this, we have counselors and placement specialists waiting for you. There's no rush. Feel free to relax a bit after your journey. When you're ready, talk with them to get started. Best of luck to each of you."

Virgil made small talk with a few people, then excused himself and drew me into a side room. "I meant that for you as well," he said. "I'd like to hear what you feel success looks like for you going forward."

I spent a moment clearing my head. My stomach was rumbling, and my brain still felt a bit disconnected. Eventually, I gathered my thoughts and began.

"I've got to take all this back to my team. I'll work with each of them individually. One on one, we can work through the issues they're encountering. I'll focus on what success on this project really looks like to each of them. Then, with those various viewpoints in mind, I can give the team a unified structure that meets all their needs, and I have you to thank for that."

"You are welcome," Virgil said. "Now let's get you out

of here and back to saving your team." He escorted me to the parking lot, where I thanked him again for his time and invitation. He smiled graciously. "You are welcome. I enjoyed seeing this from an outsider's perspective. Thank you for the time and effort you put in with me today. I'd like to leave you with one last piece to reflect on. It comes from one of my inspirations for building this school. *Di sua maggior magagna / conosce il danno; e però non s'ammiri / ce ne riprende, perché men si piagna.*

"Loosely translated, it means, 'Of his own greatest blemish he recognizes the bane; and heretofore let there be no wonder if he reproves it, that he may the less mourn for it.' I take it to mean, 'Find your fault, own the fact that it's a real problem, and find a way to fix it and do better.' It's at the heart of what we do here. Every student must do this to succeed. And they're better people because of it."

"I hope I can live up to that," I replied.

"When we started, I promised that you'd get a chance to ask questions at the end. Here's your opportunity. What have you got?" Virgil asked.

"Well, what's next for you?"

Virgil's eyes betrayed a twinkle. "I've already started thinking about how I could design a brain-computer interface that would allow me to virtualize my personality and knowledge so they would be available to many students at once. An old friend of mine has been doing some interesting work on digital reconstruction of memory states, and I think I may just have to give him a call. Thank you for coming today. I think we both found new paths to explore."

"Thanks again for everything."

I got into my car, and with my taillights illuminating the institution, I left data purgatory behind. When I got home, I opened my email to find 394 unread emails from the day. I clicked into my draft email folder, opened the "Project Canceled" draft, and deleted it. I felt a weight lift from my shoulders. I started a new draft and addressed it to my team and the CTO. The first line read, "First, let's agree on the specific problem we need to solve."

ABOUT THE AUTHOR

Andrew Barwis is a writer, software engineer, and Dante enthusiast. Originally from Milwaukee, Wisconsin, Andrew now lives in Minneapolis, Minnesota, where he solves technical problems from a deep underground bunker. He works at every scale, from FAANG companies down to the smallest startup. On the rare occasion that he makes it above ground, Andrew enjoys biking, canoeing, and dabbling in his wood shop. He believes a good question is worth a dozen good answers and in data-driven decision-making.

Made in the USA
Monee, IL
25 September 2023

42809019R00050